AI: ALIVE OR NOT?

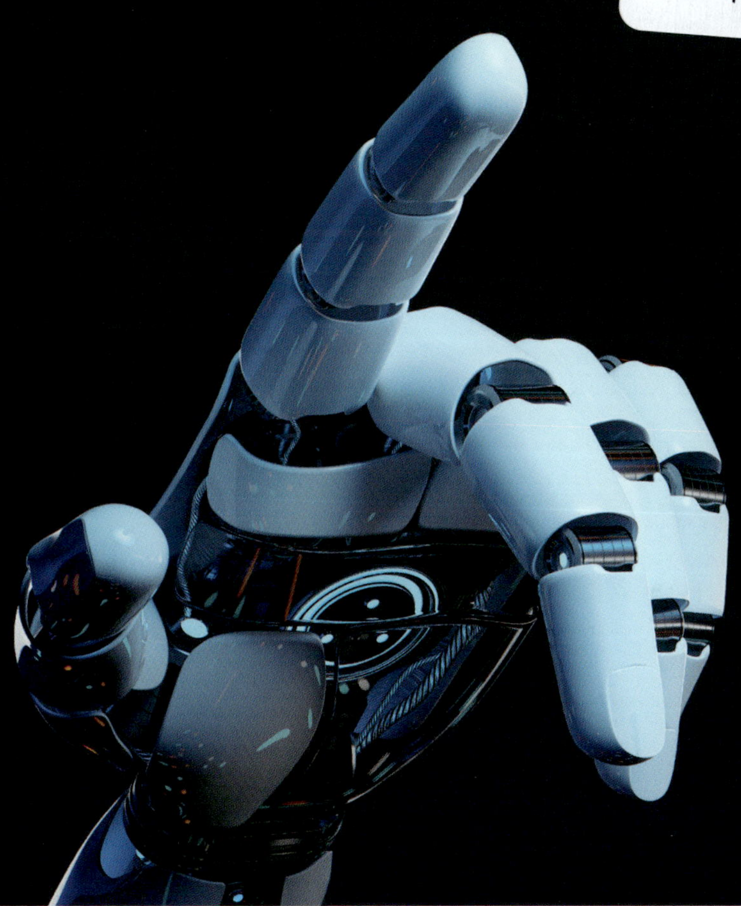

by Josh Gregory

Cherry Lake Press
Ann Arbor, Michigan

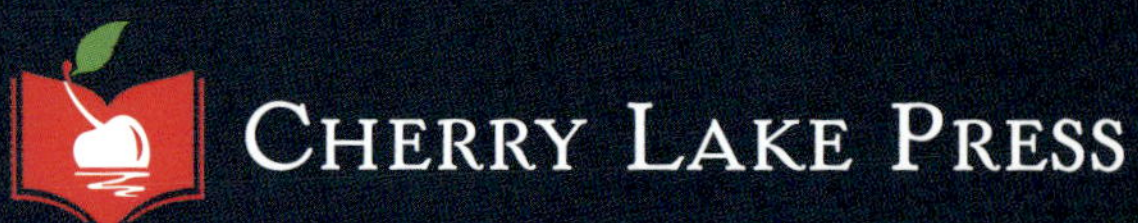

Published in the United States of America by Cherry Lake Publishing
Ann Arbor, Michigan
www.cherrylakepublishing.com

Reading Adviser: Beth Walker Gambro, MS, Ed., Reading Consultant, Yorkville, IL

Photo Credits: © sdecoret/Shutterstock, cover, title page; Matthew J. Cotter from Wigan, United Kingdom, CC BY-SA 2.0 via Wikimedia Commons, 5; © phol_66/Shutterstock, 6; © Crystal Eye Studio/Shutterstock, 9; © Gorodenkoff/Shutterstock, 11; DALL·E with GPT-4, Public domain, via Wikimedia Commons, 13; © Jo Panuwat D/Shutterstock, 14; © Stock-Asso/Shutterstock, 17; © Gorodenkoff/Shutterstock, 19; © Rawpixel.com/Shutterstock, 21; © Gorodenkoff/Shutterstock, 25; © Iurii Motov/Shutterstock, 26; © Sutthiphong Chandaeng/Shutterstock, 27; GoodStudio/Shutterstock, 28; © Fractal Pictures/Shutterstock, 29

Cherry Lake Press is an imprint of Cherry Lake Publishing Group.

Library of Congress Cataloging-in-Publication Data has been filed and is available at catalog.loc.gov.

Cherry Lake Press would like to acknowledge the work of the Partnership for 21st Century Learning, a Network of Battelle for Kids. Please visit Battelle for Kids online for more information.

Printed in the United States of America

ABOUT THE AUTHOR

Josh Gregory is the author of more than 200 books for kids. He has written about everything from animals to technology to history. A graduate of the University of Missouri–Columbia, he currently lives in Chicago, Illinois.

CONTENTS

Chapter 1 **Thinking Machines** 4
Chapter 2 **Learning and Changing** 10
Chapter 3 **What Is "Alive"?** 16
Chapter 4 **Looking Into the Future** 24

Activity 30
Find Out More 31
Glossary 32
Index 32

Chapter 1

THINKING MACHINES

Have you seen the famous old movie *2001: A Space Odyssey*? It has a character called HAL 9000. HAL is a computer that thinks and talks like a human. HAL is on a space mission with human crew members. The humans start to wonder if HAL is working correctly. They discuss unplugging HAL. But HAL overhears them. It doesn't want to be shut down. So it decides to attack the humans.

This movie was made in the 1960s. Computers were still new technology. They were unlike anything people had seen before. But people already wondered how these machines would change in the future. Even very early computers could do complicated math problems faster than humans. What if they became smarter than people in other ways? What if they learned to think for themselves? What if they became fully alive? Would they help humans? Or would they fight against us and take over the world?

Spacesuits from *2001: A Space Odyssey*. Though this movie was made in the 1960s, the fears the characters had about AI are still relevant today.

Countless movies and TV shows explore these kinds of questions about computers. Science fiction authors wrote books about living machines before computers were even invented. Today, Marvel movies have characters like Vision. Vision has a robot body. His mind is a computer program. But he thinks for himself. He becomes friends with other heroes. He even falls in love.

Writers and filmmakers are not the only people thinking about bringing computers to life. Inventors and researchers have been trying to make these stories come true for decades. They have tried to create artificial intelligence (AI) since the first computers were built. At first, computers were not powerful enough to create AI. But this is starting to change.

Vision is an **android**. Over time, Vision tries to be more "human." He falls in love with Wanda Maximoff, the Scarlet Witch.

AI technology has made huge advances. And some of it is starting to seem like it came right out of an old science fiction movie.

Have you had a conversation with ChatGPT? This chat program can write answers to any question you ask. Its replies usually make a lot of sense. It can understand all kinds of questions. It seems to know about a lot of things. It can even seem a little like chatting with a human. Is it alive?

Or think about self-driving cars and other robots that move on their own. They use cameras and other sensors to notice things around them. They use this information to make decisions. They move around in the world without following a set path. This is a lot like the way people sense the world and move around. Are these machines alive?

Experts debate whether artificial intelligence will ever stop being artificial. Will machines be able to think like humans? Can they already? The debate is changing all the time. New technology is on the way. It can be a lot to follow. But it is an important part of our world today. And it will be even more important in the future.

FROM SCI-FI TO REALITY

Authors have written science fiction stories for hundreds of years. They dream up technology that doesn't yet exist. They imagine situations that could happen one day. Characters in these stories travel through time and meet aliens. They visit faraway planets and make important discoveries. Some science fiction authors get their ideas from real-life science and technology. Others let their imaginations run free.

Many ideas in science fiction stories have come true. Authors wrote books about space travel long before the first satellite was launched. The novel *Somnium* was written in the 1600s. In it, characters visit the moon! Other authors made predictions about the internet and AI. Sometimes these authors simply make good predictions. But sometimes their books become reality because scientists and inventors read them. Then they try to bring the ideas from the books to life!

BB-8 is a droid in the *Star Wars* universe. *Star Wars* characters travel to different planets and meet new alien species. They are always discovering new things!

Chapter 2

LEARNING AND CHANGING

AI might seem like brand-new technology. Many new kinds of AI have been invented in recent years. And AI technology is quickly getting more powerful. But the basic ideas behind AI have been around for a long time.

AI is simply a type of computer program. It works differently from regular computer programs. Regular programs have a series of steps for a computer to follow. Computer programmers write these steps using **code**. Computers follow the steps exactly. They do the same thing every time. The program does not get better or worse. It can only change if programmers rewrite its code.

AI programs are more complex. They are designed to work like human minds. This means they can learn. They can get better at things. Programmers still write code

Computer programmers are still the ones who design AI programs. But they design the programs to be able to learn and improve on their own.

for AI programs. But this is just the start. Their code tells a computer how to study **data**. Data can be anything from a list of numbers to a photo. **Developers** give their AI programs a lot of data to study. The AI programs look for patterns in this data.

The basic ideas behind machine learning were developed in the 1940s. Researchers came up with the idea to build **neural networks**. These computer systems are based on the human brain. They are made up of sections called nodes. Nodes are connected. Information can move between them.

People tried to build neural networks. But computers weren't powerful enough to automate the math involved. This fact changed only recently. Today's computer systems are very fast and powerful. Neural networks have become common. Developers can use them to create AI programs.

Today, many kinds of AI programs are available. Most of them are designed to do one thing. For example, the DALL-E AI creates images. Users type out what they want to see. DALL-E creates an image based on this written text. The AI was trained using millions of images.

This image was made using DALL-E.
The text prompt was "A modern architectural building with large glass windows, situated on a cliff overlooking a serene ocean at sunset."

It learned how words are used to describe different things in images.

More advanced AI systems combine different programs. Some experts believe these AI systems can one day be as complex as a human mind. Each part of the system will do a different thing. The parts will work together to perform complex tasks. This is how a human mind works. Part of your brain controls your vision. Another deals with words. Another helps you move your body. You combine all of these—and others—to write words.

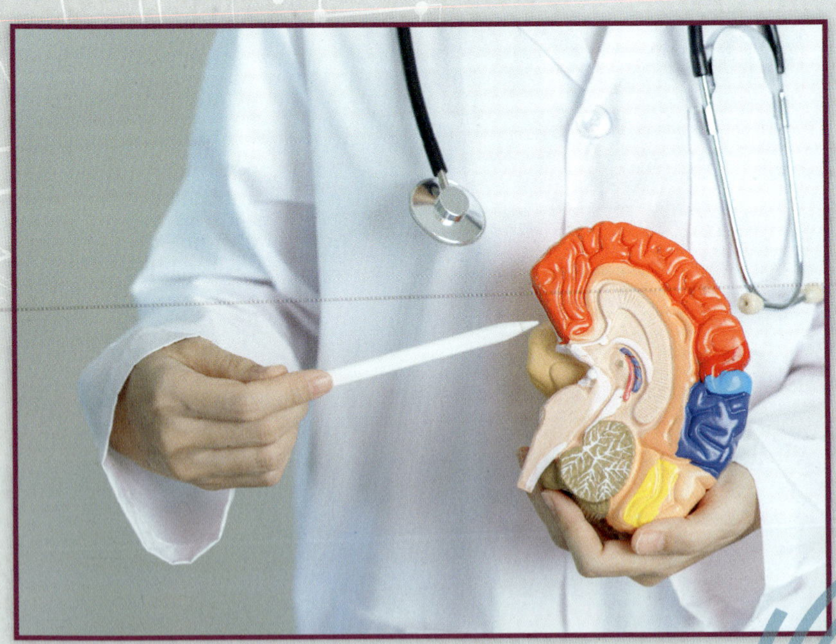

Different parts of the brain do different things. The occipital lobes help you process what you see. The temporal lobes mostly process what you hear.

THE FATHER OF AI

One of the first people to think about AI was Alan Turing. Turing was a math expert. In the 1930s, he helped invent some of the earliest computers. These computers were used to break coded messages during World War II (1939–1945).

Turing believed computers were similar to human brains. He thought computers would one day be able to think like people. In 1950, he proposed a test that could be used to decide if a computer was able to think on its own. The Turing Test is still an important idea today. Researchers talk about it when debating whether AI programs are truly "thinking" or just following instructions.

Chapter 3

WHAT IS "ALIVE"?

What exactly does it mean for something to be alive? A dog can't answer questions like ChatGPT. A fish can't solve math problems. A spider can't identify different people in photos. But everyone agrees that animals are living things. Or consider a tree. It is a living thing. But it can't think or move like an animal can. Incredibly tiny **bacteria** are alive too. And they are nothing at all like humans.

Humans have debated the meaning of life for thousands of years. Some of today's scientists agree that all life has certain things in common. Most living things can grow. And they can reproduce.

AIs are nowhere near meeting this definition of life. But this isn't what most people are thinking about when they wonder if AIs can be alive. They are wondering

If artificial intelligence programs could experience emotion, would that be helpful? What if they could feel compassion? What if they could feel anger?

if AIs can be self-aware and **conscious**. This would mean AIs know they are AIs. They would know that they are able to think. They would have their own thoughts and opinions. They'd be able to make decisions. They might have feelings. They would react to things in their own way.

Right now, none of this is true. And it is unlikely to be true anytime soon. AI programs can often seem like they are thinking for themselves. They can have conversations. They can do surprising things that even their creators don't predict. They can defeat expert chess players. They can drive cars. But none of this is proof of being self-aware. These AI programs are simply doing what they've been told to do.

Some people believe that creativity is another major part of thinking like a human. AIs can seem creative. They can make photos and videos. They can come up with poems and songs. But many people think the things they make aren't truly creative. Some people think true creativity means coming up with something original.

Many believe that AI chess programs are great training tools for humans. They can analyze how well a player will do. They can help people see what new strategies will help their game.

AIs cannot make completely new things. Everything they make is based on the data they've learned from. A music-creating AI can combine different ideas from songs it studies. But it cannot create a whole new style from scratch. ChatGPT can imitate different writing styles. But it cannot come up with a new one.

Humans often reuse others' ideas too. They copy each other. They combine each other's ideas in new ways. But human taste and perception come into play here. People are more likely to remember ideas they connect with emotionally. AIs cannot do this.

ASK AN AI

Let's go straight to the source and ask the AI text generator ChatGPT what it thinks about AIs being alive.

Hi, ChatGPT
Hello! How can I assist you today?

Are you alive?
No, I'm not alive. I'm a computer program designed to help with information and tasks.

Will AIs ever be able to think for themselves?
Unlikely. AI lacks consciousness, and current technology doesn't suggest it will gain true self-awareness.

Are any researchers trying to build self-aware AIs?
Some researchers explore AI with increased awareness, but achieving true self-awareness remains a complex challenge.

Are there any risks in building self-aware AIs?
Potential risks include ethical concerns, loss of control, unintended behaviors, and the impact on society. Careful development is crucial to **mitigate** these risks.

Thanks for sharing your thoughts!
You're welcome! If you have any more questions or if there's anything else I can help you with, feel free to ask. Have a great day!

(This conversation has been slightly edited and condensed for space.)

Chapter 4

LOOKING INTO THE FUTURE

Today's AI programs are not alive. They are not conscious. They are simply computer programs. It's possible they may never be any more than that. AI may never have a life of its own. But we don't know what the future holds. Technology is always changing. Many things we now use every day were once thought impossible.

Self-aware AI programs would change the world. They could act against their creators' wishes. They would have their own goals. These could be very different from human goals. AIs could look for ways to make themselves more powerful. They might even be able to make their own new AIs.

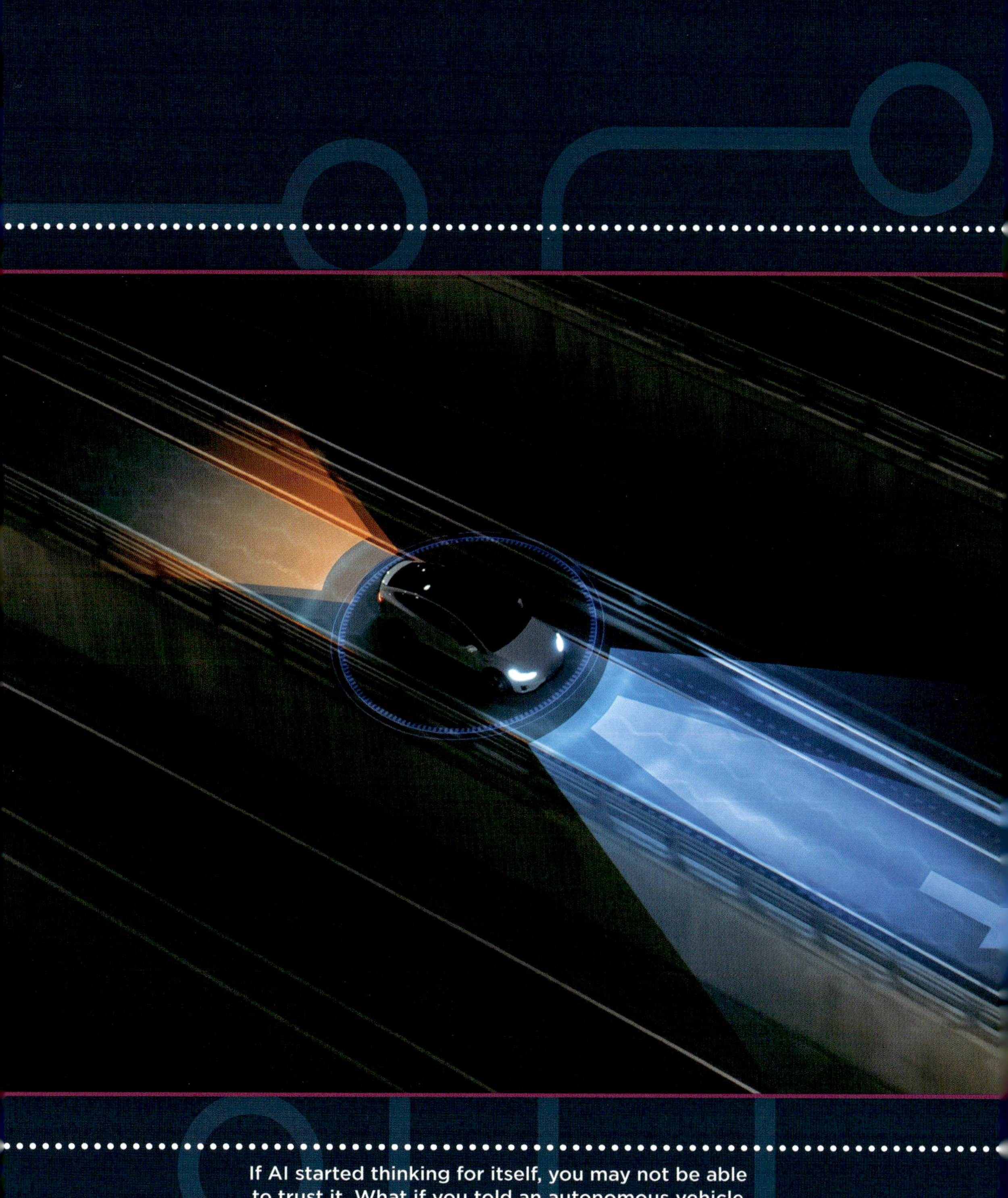

If AI started thinking for itself, you may not be able to trust it. What if you told an autonomous vehicle to go one place, but it took you to another?

We would have to decide how to treat these new AIs. Would they have rights? Imagine a computer that is alive. It has its own thoughts and dreams. Would it be OK to turn it off? Or would that be like murder? Would AIs have to follow human laws? How could we make sure they followed the rules? Imagine that your phone simply stopped working because it wanted to. Or imagine that it lied to you when you asked for information.

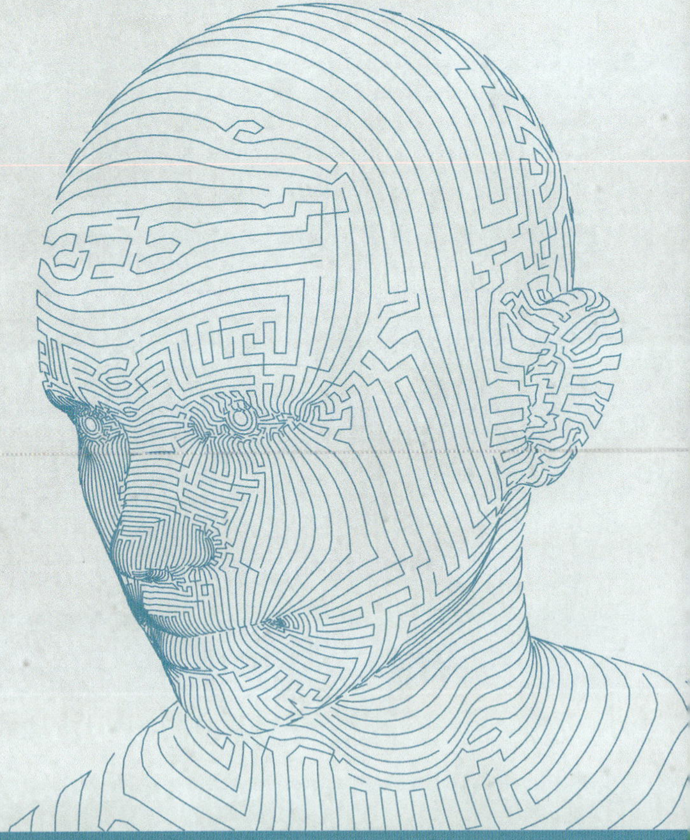

AI is helpful. But people are questioning if the disadvantages of AI will outweigh its advantages.

Many people wonder whether we should even try to create more powerful AIs. These AIs might help solve important scientific problems. They might power useful technology. They could make our lives easier. But they could also cause a lot of problems.

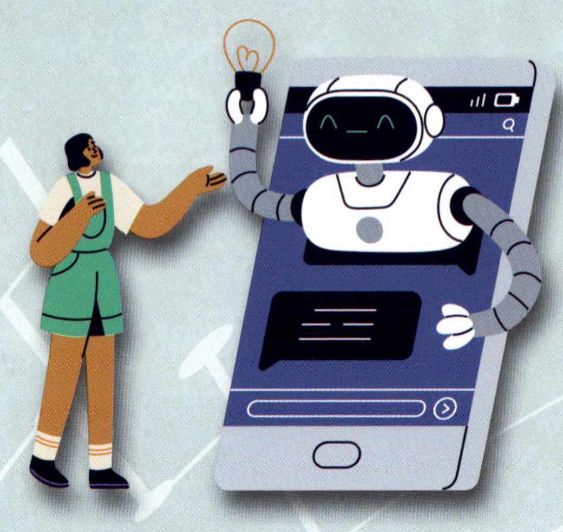

No one knows exactly what will happen with AI technology. It looks likely to become much more powerful. But it might still never think for itself. It is worth thinking about this idea, though. People have wondered about it for a very long time. And they will probably keep wondering far into the future.

LIVING FOREVER

Some companies are using AI to create copies of real people. Someone can sign up to be interviewed. The interviews are recorded. Then they are used to train an AI. People can chat with the AI. It will answer like the person who was interviewed. It has their voice. And it says the kinds of things it thinks they would say.

This service is meant for people who want to leave an AI-powered memory of themselves behind when they die. Some families think this is a great way to remember relatives. Others find it creepy. The AI is similar to the dead person. But it is not the same thing.

ACTIVITY: IMAGINING THE FUTURE

How do you think AI will change in the future? Write your own sci-fi story to explain. Here are some things to think about as you write:

- When will your story take place? In the far future? Or closer to the present? How else will the world be different at that time? Include details in your story.
- Will AIs in your story be self-aware? What will human characters think of them?
- Will there be new laws against AIs?
- What other kinds of new technology might exist in the future? How will it affect your story?

These are just ideas to get started. Be creative. Anything you imagine can go in your story!

FIND OUT MORE

Books

Abell, Tracy. *Artificial Intelligence Ethics and Debates.* Lake Elmo, MN: Focus Readers, 2020.

Felix, Rebecca. *Artificial Intelligence: Can Computers Take Over?* Minneapolis, MN: Checkerboard Library, 2019.

Gregory, Josh. *Drones.* Ann Arbor, MI: Cherry Lake Publishing, 2017.

Nadin, Joanna. *Alan Turing.* New York, NY: Scholastic, 2020.

On the Web

Search these online sources with an adult:

"Artificial Intelligence." Britannica for Kids.

"ChatGPT." OpenAI.

"Could a robot become president?" National Geographic for Kids.

"Science fiction facts for kids." Kiddle.

"What is artificial intelligence (AI)?" IBM.

GLOSSARY

android (AN-droyd)
a robot with a human-like appearance

bacteria (bak-TEER-ee-uh)
tiny living things that exist everywhere but are too small to see without a microscope

code (KOHD)
instructions written in computer programming language

conscious (KAHN-shuhs)
aware of something

data (DAY-tuh)
information used to create, process, or support something

developers (dih-VEH-luh-purz)
people who create AI and other computer programs

mitigate (MIH-tuh-gayt)
to make less harsh by managing

neural networks (NER-uhl NET-werks)
computer systems built to work like human brains

INDEX

aliveness, 16–23
androids, 6, 9
autonomous vehicles, 7, 25

bacteria, 16
BB-8, 9
brain, 12, 14

cars, self-driving, 7, 25
ChatGPT, 7, 16, 20, 22–23
chess, 18, 19
code, 10–12
computer programmers, 10–12
computer programming, 10–14
consciousness, 18, 22–23, 24
creativity, 18–20

DALL-E, 12–14
data, 12
death, 29
developers, 12

emotions, 17, 18
ethical concerns, 23, 24–27

future of AI, 24–30

HAL 9000, 4

learning, machine, 10–14
living things and aliveness, 16–23

machine learning, 10–14

neural networks, 12
nodes, 12

robots, 6, 7, 17

science fiction, 4-9, 30
self-awareness, 18, 22–23, 24
self-driving cars, 7, 25
Somnium (Kepler), 8
Star Wars, 9

thinking, 4, 7, 15, 16–18, 20, 28
Turing, Alan, and Turing Test, 15
2001: A Space Odyssey, 4

vehicles, autonomous, 7, 25
viruses, 16
Vision (Marvel character), 6